UNDER THE SEA ANIMALS

MANATEES

Amy Culliford

TABLE OF CONTENTS

Sight Words.......................................2
Words to Know.....................................3
Index...16

A Pelican Book

Teaching Tips for Caregivers and Teachers:

Research shows that one of the best ways for students to learn a new topic is to read about it.

Before Reading

- Read the title and predict what the book will be about.
- Read the "Words to Know" and discuss the meaning of each word.
- Read the back cover to see what the book is about.

During Reading

- When a student gets to a word that is unknown, ask them to look at the rest of the sentence to find clues to help with the meaning of the unknown word.
- Motivate students with praise and encouragement.

After Reading

- Discuss the main idea of the book.
- Ask students to give one detail that they learned in the book.

Sight Words

a
all
big
eat
have
is
live
many
some
this
under
water

Words to Know

flippers

manatee

noses

seagrass

tails

This is a **manatee**.
manatee

All manatees live under water.

Many manatees eat **seagrass**.

seagrass

Some manatees have big **noses**.

nose

Some manatees have big **tails**.

tail

All manatees have **flippers.**

flippers

Index

big 10, 12
eat 8
flippers 14, 15
live 6
manatee 4, 6, 8, 10, 12, 14
tails 12, 13
water 6

Written by: Amy Culliford
Design by: Under the Oaks Media
Series Development: James Earley
Editor: Kim Thompson

Photos: Jeff Stamer: cover; Bildagentur Zoonar Gmbtt: p. 4-5; Thierry Eidenwell: p. 7, 14-15; studiaprofi: p. 9; gary powell: p. 11; Ethan Daniel: p. 13

Library of Congress PCN Data
Manatees / Amy Culliford
Under the Sea Animals
ISBN 978-1-63897-065-1 (hard cover)
ISBN 978-1-63897-151-1 (paperback)
ISBN 978-1-63897-237-2 (EPUB)
ISBN 978-1-63897-323-2 (eBook)
Library of Congress Control Number: 2021945228

Printed in the United States of America.

Seahorse Publishing Company
www.seahorsepub.com

Published in the United States
Seahorse Publishing
PO Box 771325
Coral Springs, FL 33077